T0224954

Cambridge Elements ☰

Elements in Geochemical Tracers in Earth System Science
edited by
Timothy Lyons
University of California, Riverside
Alexandra Turchyn
University of Cambridge
Chris Reinhard
Georgia Institute of Technology

TRIPLE OXYGEN ISOTOPES

Huiming Bao
Peking University and Louisiana State University

CAMBRIDGE
UNIVERSITY PRESS

CAMBRIDGE
UNIVERSITY PRESS

University Printing House, Cambridge CB2 8BS, United Kingdom

One Liberty Plaza, 20th Floor, New York, NY 10006, USA

477 Williamstown Road, Port Melbourne, VIC 3207, Australia

314–321, 3rd Floor, Plot 3, Splendor Forum, Jasola District Centre,
New Delhi – 110025, India

79 Anson Road, #06–04/06, Singapore 079906

Cambridge University Press is part of the University of Cambridge.

It furthers the University's mission by disseminating knowledge in the pursuit of
education, learning, and research at the highest international levels of excellence.

www.cambridge.org
Information on this title: www.cambridge.org/9781108723374
DOI: 10.1017/9781108688543

© Huiming Bao 2019

First published 2019

A catalogue record for this publication is available from the British Library.

ISBN 978-1-108-72337-4 Paperback
ISSN 2515-7027 (online)
ISSN 2515-6454 (print)

Triple Oxygen Isotopes

Elements in Geochemical Tracers in Earth System Science

DOI: 10.1017/9781108688543
First published online: August 2019

Huiming Bao
Peking University and Louisiana State University
Author for correspondence: Huiming Bao, bao@lsu.edu

Abstract: Since the first discovery of isotopes of elements 100 years ago, the "detective" power of stable isotopes for processes that occurred in the past, and for elucidating mechanisms at the molecular level, has impressed researchers. While most are interested in the normalized abundance ratios of two isotopes of an element, further power was unleashed when researchers investigated the relationship of three or more isotopes of the same element, e.g., ^{16}O, ^{17}O, and ^{18}O for oxygen. This Element focuses on the history of discovery of triple isotope effects, the conceptual framework behind these effects, and major lines of development in the past few years of triple oxygen isotope research.

Keywords: Ozone, ^{17}O anomaly, Triple isotope exponent, Slope, Minor deviations, δ-δ space, 0.5305

ISBNs: 9781108723374 (PB), 9781108688543 (OC)
ISSNs: 2515-7027 (online), 2515-6454 (print)

Contents

1. Introduction

Research into triple oxygen isotopes studies the relationship between the ratio of $^{18}O/^{16}O$ and that of $^{17}O/^{16}O$ during an elementary process or a set of processes. The processes can be physical, chemical, and/or biological, and can be specifically or loosely defined.

J. J. Thompson found that a parabola of neon ions at the atomic weight of 20 on a photographic plate was accompanied by a line with an atomic weight of about 22 when neon ions passed through parallel electric and magnetic fields. As he concluded in his Bakerian Lecture at the Royal Society in 1913 (Thompson, 1913), "There can, therefore, I think, be little doubt that what has been called neon is not a simple gas but a mixture of two gases, one of which has an atomic weight about 20 and the other about 22. The parabola due to the heavier gas is always much fainter than that due to the lighter, so that probably the heavier gas forms only a small percentage of the mixture." The two isotopes of neon were later established by F. W. Aston using a much-improved positive-ray spectrograph or "mass spectrograph" he built and named (Aston, 1919b, c, 1920b), thus confirming his suspicion of neon "leading a double life" (Hughes, 2009). What Aston also suspected at that time was that neon possibly had a third isotope, ^{21}Ne, but his spectrograph did not give him the confidence to declare this. However, Aston did state that if ^{21}Ne existed in the atmospheric sample, its abundance would be "probably well under 1%" (Aston, 1920a). The third isotope of neon, ^{21}Ne, was confirmed later by a group at the University of California (Hogness and Kvalnes, 1928), with the lowest abundance, estimated to be at 2 percent, among the three (we know today that ^{21}Ne is 0.27 percent in abundance). That said, the first element confidently reported to have three and probably more isotopes was mercury, at the very beginning of Aston's search for isotopes (Aston, 1919a).

Oxygen was also measured by Aston and was labeled a "pure element", i.e., no second isotope (Aston, 1920b). Minor isotopes of oxygen, ^{18}O and ^{17}O, in Earth's atmosphere were later discovered spectroscopically (Giauque and Johnston, 1929a, b). Aston was not convinced initially, and his mass spectrograph let him conclude that even if ^{18}O existed, its abundance must be less than 1:1000 (Aston, 1929). A few years later Aston made a more accurate mass-spectrum determination of ^{16}O, ^{17}O, and ^{18}O abundances (Aston, 1932).

Early efforts in stable isotope research, driven by an interest in nuclear physics (Hughes, 2009), focused on the discovery of new isotopes, which in turn fueled the rapid development of nuclear physics after 1930 (Brickwedde, 1982). The application of stable isotopes in geology came much later, after the statistical mechanic ground for isotope effect was established in 1947 by Urey,

Bigeleisen, and Goeppert-Mayer (Bigeleisen and Mayer, 1947; Urey, 1947), as an outcome of the Manhattan Project (Bigeleisen, 2006). The utilities of simultaneously measuring all three stable oxygen isotopes were not explored until 1973 when Robert N. Clayton's group at the University of Chicago discovered an unusual variation of ^{17}O abundance in meteorites (Clayton et al., 1973). Since then, triple oxygen isotope data have become indispensable in studying extraterrestrial materials, e.g., classifying meteorites. In 1983, however, this once exclusively "celestial" signature was brought down to Earth in the laboratory, and later in atmospheric ozone (O_3) by Mark Thiemens' group at the University of California in San Diego (Johnston and Thiemens, 1997; Thiemens and Heidenreich, 1983). In the subsequent years, many oxygen-bearing Earth atmospheric compounds were found to have variable degrees of anomalous ^{17}O composition that do not fall on the terrestrial fractionation line (Thiemens, 1999). Non-mass-dependent (NMD) or mass-independent (MIF) processes or signatures were suspected to be involved in the formation or destruction of these atmospheric compounds. Note that neither "NMD" nor "MIF" is literally accurate in describing the process or signature that deviates from the classical Urey–Bigeleisen–Mayer mass fractionation law because most of the fractionation is still the result of mass difference and its consequences. In 2000, NMD ^{17}O signatures were first discovered in Earth's rock records (Bao et al., 2000b), bringing the anomalous signature further down to Earth and its distant past. The year 2008 marks the first discovery of anomalously ^{17}O-depleted compositions in the rock record (Bao et al., 2008). This is significant because, between 2000 and 2008, all cases of terrestrial ^{17}O anomalies were ^{17}O enrichment when compared to the oxygen of bulk silicate Earth or seawater. We have recognized that all these discovered ^{17}O enrichments can be directly or indirectly traced back to tropospheric and stratospheric O_3, the triple oxygen isotope composition of which is highly and anomalously enriched in ^{17}O. The anomalous ^{17}O depletion in sulfate minerals of ~635 million years ago (Ma) is proposed to originate directly from atmospheric O_2, the anomalous ^{17}O depletion of which is, nevertheless, indirectly linked to stratospheric O_3 (Bao et al., 2008; Cao and Bao, 2013). In fact, anomalous ^{17}O depletion was discovered and a mechanism was proposed for modern atmospheric O_2 in 1999 (Luz et al., 1999). The 635 Ma case is simply an extreme version of atmospheric composition and chemistry in which the CO_2 concentration was ultra-high, at a concentration level equivalent to O_2, while photosynthetic O_2 input was at the modern level.

The brief history of triple oxygen isotope research outlined so far was focused on ^{17}O anomalies that are large in magnitude, and probably all originated from chemistries related to atmospheric O_3 formation and destruction. In recent years, with analytical improvements, minor deviations of ^{17}O content from a triple oxygen isotope reference frame became resolvable. These minor deviations can result from slight differences in triple oxygen isotope relationships among different, yet nominally mass-dependent, processes. Because of the ubiquitous presence of oxygen in solids, liquids, and gases, this development has forged a path to study lithosphere, hydrosphere, atmosphere, and biosphere processes, and their interactions in different space and time scales.

Much of the basic concepts, analytical issues, and major applications on triple oxygen isotopes can be found in a 2016 review paper (Bao et al., 2016). After recounting the processes leading to the first measurement of the triple oxygen isotope composition, I will go over the key concepts pertinent to study of triple oxygen isotopes, explore the origin of large and minor ^{17}O anomalies, respectively, summarize analytical techniques, and present the latest developments in triple oxygen isotopes research. A list of representative papers on the theories, original discoveries and applications, and reviews is annotated at the end.

2. The First Measurement of $\delta^{17}O$ Value

Blackett first detected ^{17}O from a nuclear reaction in 1924 (Blackett, 1925). Since Aston's first mass spectrometry confirmation of ^{16}O, ^{17}O, and ^{18}O abundance in 1932 (Aston, 1932), many others have attempted to improve measurement of the relative abundances (Murphey, 1941). The first measurement of $\delta^{17}O$ value on a natural compound was only possible after a reference was designated for oxygen isotope ratios in 1961 (Craig, 1961).

There are many cases in the history of stable isotope geochemistry in which an initial research plan led to a totally unexpected, yet much more impactful discovery. Here is another. In the early 1970s, equipped with a theoretical understanding of oxygen isotope geothermometry and available meteorite samples, the Chicago group led by Robert N. Clayton analyzed oxygen isotope compositions of meteorites. They were aiming to discover the temperature at which minerals in the meteorites formed in solar nebula gas. As Clayton recalled (Clayton, 2007; 2008), Larry Grossman, then a new assistant professor who had just attained his PhD degree from Yale University, was working on the Allende meteorite in which he discovered abundant calcium-aluminum-rich inclusions (CAIs). CAIs are regarded as high-temperature condensates, and, therefore, are a good target for extending geothermometer to extraterrestrial conditions. At that time, all oxygen isotope composition was run using CO_2 gas by measuring the mass of 44, 45, and 46. Even if silicate or oxide minerals were converted to O_2 gas via fluorination, the O_2 was passed through red-hot graphite to turn O_2 into CO_2 for measurement on a mass spectrometer. However, if the $\delta^{17}O$ value were measured by running as CO_2, the precision would be magnitudes worse than running as O_2. This is because both isotopologues, $^{13}C^{16}O^{16}O$ and $^{12}C^{16}O^{17}O$, will be registered in faraday cup 45, and both $^{12}C^{16}O^{18}O$ and $^{13}C^{16}O^{17}O$ in cup 46. These isobaric interferences in the value of $\delta^{18}O$ are not significant, and only a small correction is needed for the corresponding $\delta^{13}C$ value as initially established by Harmon Craig (1957). However, the mass contribution of $^{13}C^{16}O^{16}O$ to cup 45 is two magnitudes larger than that of $^{12}C^{16}O^{17}O$, rendering the accuracy of the $\delta^{17}O$ of CO_2 unlikely even if pure ^{12}C graphite was used. At that time, the practice of O_2 to CO_2 conversion before mass spectrometer analysis made sense because the references or standardization practices among laboratories had been based on CO_2, the $\delta^{18}O$ measurement using CO_2 was accurate, and there was little interest in $\delta^{17}O$.

Because the Chicago group's O_2 samples of different $^{18}O/^{16}O$ ratios all went through the same graphite (thus the same $\delta^{13}C$) before analysis, when plotted on space defined by 45/44 ratio and 46/44 ratio, these data points displayed a slope of 0.034, as seen repeatedly among minerals from the Earth (terrestrial). The slope value of 0.034 is essentially the terrestrial molar abundance ratio of $^{17}O/^{13}C$. However, after many runs of the meteorite samples, Clayton's team noted that the slope was not at 0.034, but nearly twice that for the measured CAIs. Immediately they knew that the abundance ratio of $^{17}O/^{13}C$ was different for CAIs, and perhaps for many extraterrestrial materials. From that point on, they switched from measuring CO_2 to O_2 generated from fluorination. They published a seminar paper in 1973 (Clayton et al., 1973), and started a new era in stable isotope cosmochemistry. Additionally, running O_2 almost avoids isobaric interferences, and both $\delta^{17}O$ and $\delta^{18}O$ values can be obtained simultaneously and accurately, not to mention that running O_2 directly also eliminates the extra O_2 + graphite burning step in the experiment.

3. Terminology and Concepts in Triple Oxygen Isotopes

3.1 Triple Oxygen Isotope Compositions

The measurement of $\delta^{17}O$ values alone does not convey much information if the associated $\delta^{18}O$ is not also measured for comparison. Thus, triple oxygen isotope composition refers, in fact, to paired $\delta^{18}O$-$\delta^{17}O$ or $\delta^{18}O$-$\Delta^{17}O$ data. Using $\delta^{18}O$-$\delta^{17}O$ space does not differ from using $\delta^{18}O$-$\Delta^{17}O$ space in demonstrating triple oxygen isotope relationships, but the latter rotates the reference axis so that the degree of ^{17}O deviation is emphasized.

This reference frame is essentially the way we calculate the $\Delta^{17}O$ value from measured $\delta^{18}O$ and $\delta^{17}O$ values. The following definition is commonly used:

$$\Delta^{17}O = \delta^{17}O - C \times \delta^{18}O. \tag{1}$$

There are a few variants in equation (1). The δ can be defined linearly as $\delta \equiv R_{sample}/R_{reference} -1$, or logarithmically as $\delta' \equiv \ln (R_{sample}/R_{reference})$ (R is the molar ratio of $^{17}O/^{16}O$ or $^{18}O/^{16}O$), while C can be any value, typically a number between 0.50 and 0.05305. For those studies dealing with significant ^{17}O anomalies, a canonical value of 0.52 is adopted for C, and it does not make much difference when different δ definitions are used. For those studies dealing with minor ^{17}O deviations or any cases when the $\Delta^{17}O$ value is small, the use of δ' and a C value of 0.5305 are recommended regardless of the compounds of interest (Bao et al., 2016), although the water and paleoclimate community has been using a C value of 0.528.

3.2 Triple Oxygen Isotope Relationship in a Process

The triple oxygen isotope relationship is process-specific. When a process is not defined, there is no triple oxygen isotope relationship. The δ (or δ') is an intensive property of an oxygen-bearing compound, as is the $\delta^{17}O$-$\delta^{18}O$ or $\Delta^{17}O$-$\delta^{18}O$ pair. There is no relationship between $\delta^{17}O$ and $\delta^{18}O$ or $\Delta^{17}O$ and $\delta^{18}O$ in a compound itself because the relationship exists only in a process. Thus, the ratio of $\delta^{17}O/\delta^{18}O$ of a compound does not mean anything by itself except when a process linked to the international reference sample Vienna Standard Mean Ocean Water (VSMOW) is implied. Three basic concepts are used to refer to a triple isotope relationship:

(1) α, the fractionation factor, defined as R_A/R_B in which R can be $^{17}O/^{16}O$ or $^{18}O/^{16}O$, and A and B are two separate phases or the same phase at two separate times.
(2) θ, the triple isotopes exponent, defined as $\ln^{17}\alpha/\ln^{18}\alpha$.
(3) S, the slope in $\delta^{17}O$-$\delta^{18}O$ or $\Delta^{17}O$-$\delta^{18}O$ space.

The three concepts have respective equilibrium, kinetic, diagnostic, and apparent varieties. At equilibrium, a defined process will have an α_{eq} and a θ_{eq} value together with an $S_{eq} = \theta_{eq}$ in $\delta^{17}O$-$\delta^{18}O$ space. When a transition state and both kinetic isotope effects, ^{17}KIE and ^{18}KIE, are known, we have an intrinsic θ_{KIE} value $= \ln^{17}KIE/\ln^{18}KIE$ and an $S_{KIE} = \theta_{KIE}$ in $\delta^{17}O$-$\delta^{18}O$ space. When a process consists of a series of elementary processes that may or may not be at equilibrium, we will have the apparent α_{apt}, θ_{apt}. When a regression line is drawn in $\delta^{17}O$-$\delta^{18}O$ or $\Delta^{17}O$-$\delta^{18}O$ space for data points with no explicit knowledge of the underlying elementary steps and their relations, the linear regression

Table 1 Concepts and their symbols in triple isotope relationships

Concept	Equilibrium (A-B)	KIE	Diagnostic	Apparent
α	α_{eq} (A-B)	KIE	α_{dgn}	α_{apt}
θ	θ_{eq}	θ_{KIE}	θ_{dgn}	θ_{apt}
S	$S_{eq}= \theta_{eq} =$ $\Delta\delta'^{17}O/\Delta\delta'^{18}O$	$S_{KIE}=\theta_{KIE}$	S_{dgn}	S_{apt}
	Intrinsic	Intrinsic	An equation linking to intrinsic parameters may be obtained	A link to intrinsic parameters is neither direct nor obvious

line will have a slope S_{apt}. Therefore, the only intrinsic and fundamental parameters are α_{eq} and KIEs in which α_{eq} can, in fact, be derived from forward and backward KIEs of a defined elementary reaction. The α_{apt} and θ_{apt} for a nonelementary process can be traced back to the intrinsic α_{eq} and KIE values, but other factors such as initial condition, changing reaction rate, the relationship between the multiple elementary steps, and the reservoir-transport effect will all come into play. When some of the nonelementary processes are well characterized, or a set of elementary processes behaves rather conservatively such as some of the biochemical reactions, we may have a relatively stable, thus "diagnostic" α and θ value for a nonelementary process. Thus, when a relationship among the data points in δ-δ space is known, we may have a diagnostic S, or S_{dgn}, as well. Table 1 compares these concepts and symbols.

3.3 The Physical-Chemistry of a Triple Oxygen Isotope Relationship

A triple oxygen isotope relationship is often depicted by a data trajectory on $\delta^{17}O$-$\delta^{18}O$ space. The relationship can be a straight line between two points or a regression line through many points. It is here that confusions may arise. If a pair of data points are linked in δ'-δ' space, the slope value of that straight line is equal to the corresponding θ value, being equilibrium, kinetic, diagnostic, or apparent. However, when more than two data points are in δ'-δ' space, the underlying process(es) must be known before we can interpret the physical meaning of the data trajectory. For example, if the data points represent triple oxygen isotope compositions of a set of residual samples in a Rayleigh

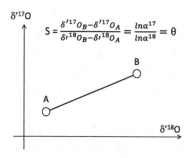 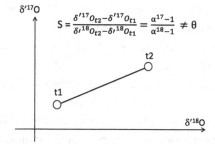

$$\delta'^{17}O \qquad S = \frac{\delta'^{17}O_B - \delta'^{17}O_A}{\delta'^{18}O_B - \delta'^{18}O_A} = \frac{\ln\alpha^{17}}{\ln\alpha^{18}} = \theta$$

$$\delta'^{17}O \qquad S = \frac{\delta'^{17}O_{t2} - \delta'^{17}O_{t1}}{\delta'^{18}O_{t2} - \delta'^{18}O_{t1}} = \frac{\alpha^{17}-1}{\alpha^{18}-1} \neq \theta$$

"A" and "B" are two phases in equilibrium, in kinetic, or in an apparent process

"t1" and "t2" are residual samples at two respective stages in a Rayleigh Distillation process

Figure 1 An example illustrating the meanings of "S" in δ'-δ' space.

Distillation process, the S_{dgn} value of the trajectory is equal to $(^{17}\alpha-1)/(^{18}\alpha-1)$, rather than $\ln^{17}\alpha/\ln^{18}\alpha$ or the θ value (Fig. 1). Although the values of these two slopes are very close, they have distinct physical meanings. For data points that have multiple origins of unknown underlying processes, the data trajectory in δ'-δ' space is neither simple mixing nor simple integration of multiple processes. In such a case, the S value has little bearing on the θ values, although they will still be close to the canonical value of 0.52 for triple oxygen isotopes.

A slope value for a data trajectory in δ-δ space does not mean much if the underlying processes are not known. However, two intrinsic parameters, θ_{eq} and θ_{KIE}, constitute the foundation of a triple isotope relationship, with a diagnostic or apparent S value in δ-δ space being influenced additionally by reservoir-transport kinetic complexities (Bao et al., 2015; Matsuhisa et al., 1978). The triple isotope relationship reveals higher dimensional information that the $^{18}O/^{16}O$ ratio alone cannot (Dauphas and Schauble, 2016).

4. Origin of Large ^{17}O Anomalies in Nature

Large degrees of NMD ^{17}O anomaly, with either positive or negative $\Delta^{17}O$ values, for terrestrial oxygen-bearing compounds can be traced to that of O_3 in the atmosphere. Although generated via different reaction pathways, O_3 in the troposphere and stratosphere both bear large, positive $\Delta^{17}O$ values.

In the troposphere, the $\Delta^{17}O$ value of O_3 ranges from ~20‰ to 40‰ (Johnston and Thiemens, 1997; Vicars and Savarino, 2014). In the stratosphere, O_3 isotope measurement has a much bigger uncertainty even for balloon-collected samples, but the $\delta^{49}O_3$ and $\delta^{50}O_3$ values (Krankowsky et al., 2007) give a $\Delta^{17}O$ range very close to that of the tropospheric O_3. The origin of the NMD ^{17}O enrichment in O_3 has been investigated extensively (Gao and Marcus, 2001; 2002; Babikov et al., 2003) since its first discovery in the laboratory (Thiemens and Heidenreich, 1983).

For the interest of Earth system history, what is pertinent is that O_3 is the most important oxidant in the atmosphere; O_3 transfers its isotopic signature to its derived oxidants such as·OH, H_2O_2, and NO_x in atmospheric processes (Lyons, 2001). As a result, O_3's NMD ^{17}O signature is transferred to its oxidation products such as sulfate, nitrate, and perchlorate (Bao, 2015; Bao and Gu, 2004; Michalski et al., 2003; Savarino et al., 2000).

O_3 concentration is much higher in part (~16–35 km) of the stratosphere than in the troposphere. Because of the dryness and the lack of biological enzymatic reaction in the stratosphere, CO_2 plays a critical role in determining the triple oxygen isotope composition of O_3, O_2, and CO_2 as they are involved in the Chapman reaction, i.e., the O_3-CO_2-O_2 reaction. Oxygen exchange among the three species occurs mainly through $O(^1D)$ produced by O_3 photolysis and via an intermediate CO_3 complex (Yung et al., 1997). The reactions result in three consequences for isotopes:

(1) Stratospheric O_3 is highly positive in both $\delta^{18}O$ and $\Delta^{17}O$, similar to tropospheric O_3. Transportation of stratospheric O_3 downward to the troposphere occurs primarily via sporadic, small-scale eddy exchange phenomena related to the global Brewer-Dobson circulation (Holton et al., 1995). The contribution of stratospheric O_3 to the troposphere pool fluctuates with season, location, and model parameters. A series of recent campaigns in North American sites revealed that, on average, the stratospheric O_3 contributions at 0–1 km, 1–3 km, and 3–8 km of the troposphere were 4.6 percent, 15 percent, and 26 percent to the O_3 budget, respectively (Tarasicka et al., 2019).

(2) Stratospheric CO_2 is also highly positive in both $\delta^{18}O$ (up to ~55‰) and $\Delta^{17}O$ (up to ~12‰) (Kawagucci et al., 2008; Thiemens et al., 1995). Once transported to the troposphere, the $\Delta^{17}O$-positive CO_2 will exchange oxygen isotopes with liquid water efficiently, resulting in a zero-$\Delta^{17}O$ value with a rather narrow variability (~0.33‰) for near-surface atmospheric CO_2, as measured recently from multiple sites around the world (Liang et al., 2017).

(3) Whereas both stratospheric O_3 and CO_2 are non-mass-dependently enriched in ^{17}O, stratospheric O_2 is correspondingly depleted in ^{17}O. Balloon measurement, showed only a slight depletion, however (Thiemens et al., 1995). Estimated stratospheric $\Delta^{17}O_{O2}$ was ~ −1.9‰ when the upper end of the measured $\Delta^{17}O$ of ~12.4‰ for stratospheric CO_2 was used (Luz et al., 1999) (numbers renormalized to VSMOW here). This stratospheric O_2 mixes with zero-$\Delta^{17}O$ O_2 derived from biospheric photosynthesis to result in modern near-surface O_2 with $\delta^{17}O$ and $\delta^{18}O$ values of 12.08‰ and 23.88‰, respectively, and thus a $\Delta'^{17}O$ value of ~ −0.51‰ (Barkan and Luz, 2005) (recalculated using C= 0.5305 here). The $\Delta^{17}O$ value of tropospheric O_2 varies

with many factors, mainly the O_2/CO_2 concentration ratio and biosphere O_2 flux (Bao et al., 2008; Cao and Bao, 2013; Luz et al., 1999). The $\Delta^{17}O$ of surface O_2 becomes increasingly negative when CO_2 concentration increases to the same magnitude as O_2 concentration and/or fresh O_2 flux from biosphere decreases and thus O_2 residence time in the atmosphere increases (Cao and Bao, 2013; Crockford et al., 2018a).

5. Origin of Minor ^{17}O Deviations

In a given reference frame, e.g., $\Delta'^{17}O = 0.5305 \times \delta'^{18}O - \delta'^{17}O$, a small deviation of the $\Delta'^{17}O$ value from 0 would occur when the θ_{eq}, θ_{dgn}, or θ_{apt} value for a process of interest is not exactly at 0.5305. Among the many mass-dependent processes, the θ values may vary with process and temperature, with the θ value deviating further from the canonical value of 0.52 when the α approaches 1.000. This has been thoroughly explored (Bao et al., 2015; Hayles et al., 2017), and, as the $\Delta'^{17}O$ value scales with the corresponding $\delta'^{18}O$ value, most mass-dependent processes will generate only small $\Delta'^{17}O$ values, thus termed minor ^{17}O deviations. Modern analytical techniques can resolve some of these minor deviations and therefore high-precision measurement of triple oxygen isotope composition is opening up new frontiers (see Section 7.3).

6. Analytical Methods and Material Preparation

As illustrated by the story of the first $\delta^{17}O$ measurement, triple oxygen isotope analysis has the least isobaric interference when run as O_2 in dual-inlet mode on a gas-source isotope-ratio mass spectrometer. Therefore, most analytical development efforts have centered on conversion of oxygen-bearing gases, ions, or solids into O_2. Table 4 of Bao et al. (2016) summarizes the diverse methods used. Although most of the methods achieved ~100 percent O_2 yield, some did not. For research on large $\Delta^{17}O$ values, partial yields are not a problem because the processes responsible for partial O_2 yield are mostly mass-dependent processes. This is the case for fluorination of sulfate (Bao and Thiemens, 2000), thermal decomposition of nitrate (Michalski et al., 2002) and perchlorate (Bao and Gu, 2004), carbonate via acid digestion plus further fluorination (Barkan and Luz, 2012; Clayton and Mayeda, 1984; Passey et al., 2014), or O_2-CO_2 exchange (Mahata et al., 2016) methods. When dealing with minor $\Delta'^{17}O$ variations, the effect of partial yield should be considered because the experimental process that generates the partial yield may not have the same θ value as the processes of interest. The issue may not be serious if an assumption can be made that all data being compared were subject to the same set of experimental conditions and have a similar range of partial O_2 yields. However, when studying the triple isotope relationship among different compounds, the θ value for a partial-yielding process must be calibrated.

Both $\Delta'^{17}O$ and $\delta'^{18}O$ values must be considered together, particularly when the focus is on minor ^{17}O deviations because once a $\Delta'-\delta'$ reference frame is given, the $\Delta'^{17}O$ value scales with the $\delta'^{18}O$. Sample purification becomes more critical when dealing with minor ^{17}O deviations. To date, many laboratories have achieved a $\Delta'^{17}O$ accuracy of 5 per meg or 0.005‰ (standard error of the mean, s.e.m.) for procedures that can generate ~100 percent O_2 yield. The accuracy suffers when the yield is partial, e.g., ±0.03‰ (1σ) for sulfate $\Delta^{17}O$.

For geological samples, diagenesis, hydrothermal alteration, or metamorphism will likely alter the original $\Delta^{17}O$ values; these alterations can only erase but cannot create a sizeable ^{17}O anomaly because the involved oxygen-bearing compounds (such as water) are normal in ^{17}O content and the processes are mass-dependent.

7. Representative Case Studies of Geological Interests

Triple oxygen isotopes have been investigated for applications in many fields, including planetary sciences, atmospheric chemistry, paleoclimate, and geochemical kinetics (e.g., mineral dissolution and crystallization). Here I briefly highlight some of the latest case studies relevant to Earth system history.

7.1 Large Positive $\Delta^{17}O$

Massive, geological deposits that bear large positive $\Delta^{17}O$ values are those of sulfate, nitrate, and perchlorate in old and hyper-arid deserts, including Antarctic Dry Valleys (Bao et al., 2000a; Bao and Marchant, 2006), the Atacama Desert (Bao, 2005; Sun et al., 2018), Central Namib Desert (Bao et al., 2001), and deserts in other arid regions. These desert surfaces are as old as 14 Ma, and long-term aridity and stability have resulted in significant accumulation of atmospheric secondary salts that carry O_3's large positive $\Delta^{17}O$ signature. The gypsum and its pseudomorphs in many of the volcanoclastic deposits of ~30 Ma in the Northern High Plains of North America have some of the most positive $\Delta^{17}O$ values, reaching as high as 6‰ (Bao et al., 2003; 2010). Their origin is still speculative partly because there is not a modern analog case, but the oxidation of volcanic SO_2 must have followed a pathway in which O_3 was the overwhelming oxidant. Positive $\Delta^{17}O$ values have not been identified from any minerals older than 40 Ma, but it is likely they were produced if atmospheric O_3 was present. The lack of records may result from difficulty of preservation.

7.2 Large Negative $\Delta^{17}O$

Bao et al. (2008) discovered that the barite crystal fans on top of the cap carbonates of the Marinoan (~635 Ma) glacial diamictites are non-mass-dependently depleted in ^{17}O in their sulfate, with $\Delta^{17}O$ values as negative as -0.7‰, whereas none of the other sulfate $\Delta^{17}O$ values measured for the last 750 million years went below -0.29‰ (analytical error 1σ smaller than ±0.03‰). Subsequent studies confirmed that this Marinoan oxygen-17 depletion (MOSD) event was global; the anomaly has so far been found in South China (Bao et al., 2008; Killingsworth et al., 2013; Peng et al., 2011), West Africa (Mauritania) (Bao et al., 2008), Svalbard (Bao et al., 2009; Benn et al., 2015), Northwestern Australia (Bao et al., 2012), Northwest Canada (Crockford et al., 2016), East Finnmark, and East-Central Brazil (Crockford et al., 2018b). The MOSD event has been interpreted as evidence for an ultra-high pCO_2 level in the immediate aftermath of Marinoan snowball meltdown when biosphere O_2 flux was shown to be high, thus providing strong supporting evidence for the Snowball Earth hypothesis (Bao et al., 2008; Cao and Bao, 2013).

A recent study reported large negative $\Delta^{17}O$ values for sulfate of evaporitic origin, but of mid-Proterozoic age of ~ 1400 Ma (Crockford et al., 2018a). This discovery is significant because, in addition to being high-profile supporting evidence for the Snowball Earth hypothesis, large negative sulfate $\Delta^{17}O$ signatures can also be generated under conditions of relatively low biosphere primary productivity when atmospheric CO_2 and O_2 concentrations are at the same magnitude, as predicted by a modeling study (Cao and Bao, 2013). It is very likely that during much of the oxygenated atmospheric history, air O_2 has had distinctively large and negative $\Delta^{17}O$ values.

Today's near-surface O_2 has a $\Delta'^{17}O$ of ~-0.51‰, and O_2 is involved in sulfide oxidation, thus the product sulfate carries a portion of its oxygen from air O_2. Therefore, riverine sulfate tends to have more negative $\Delta^{17}O$ values than many other sources of sulfate. This measurable difference can be useful in studies of continental weathering intensity and dynamics (Killingsworth et al., 2018) as well as microbial sulfur redox cycles in oceans.

7.3 Minor ^{17}O Deviations

Young et al. (2002) alerted the community that minor deviations or nonzero $\Delta^{17}O$ values can be generated for mass-dependent processes when slightly variable triple isotope exponents are evaluated against a given reference frame. With improvements in analytical accuracy, further examination of minor deviations has become possible in triple oxygen (e.g., Barkan and Luz, 2007) and quadruple sulfur isotope systems (Farquhar et al., 2003). In the last few years, a range of ~0.3‰ has been found among terrestrial materials including sedimentary silicate and oxide minerals (Levin et al., 2014), mantle rocks (Pack and Herwartz, 2014), and water (Luz and Barkan, 2010). This is significant considering the analytical accuracy for $\Delta^{17}O$ is at 0.005‰ (s.e.m.), which has allowed reconstruction of relative air humidity in the recent past (Gázquez et al., 2018; Evans et al., 2018). Triple oxygen isotope compositions of silicates record extra-dimensional information for water from which the minerals interacted in the distant past (Bindeman et al., 2018; Herwartz et al., 2015; Herwartz et al., 2017); they can also serve as an independent geotherm-ometer because the θ value has a certain degree of temperature sensitivity, especially for the quartz-water system (Bao et al., 2016; Hayles et al., 2017; Hayles et al., 2018; Sharp et al., 2016; Wostbrock et al., 2018).

8. Outlook

The most fundamental parameters in triple oxygen isotope research are ^{17}KIE and ^{18}KIE for an elementary reaction. From the two, a θ_{KIE} value can be obtained. Complex processes can be reduced to fewer intrinsic variables. However, such calibration efforts have been rare by laboratory experiment, observation, or by first-principle computation. Most studies have been at the level of "diagnostic" and "apparent." Part of the reason for this is the difficulty in obtaining process-specific ^{17}KIE and ^{18}KIE. Often, we must assume, or we hope that the process of interest, despite being an integration of a set of elementary processes, is con-servative and therefore diagnostic. This approach is certainly a necessary com-promise with the current state of knowledge. However, we should bear in mind that the assumption we are forced to bear may well be wrong.

The ultimate source of the large ^{17}O anomalies in nature is atmospheric O_3 chemistry, particularly the stratospheric O_3-CO_2-O_2 reaction network. Our understanding of this reaction and its manifestation on triple oxygen isotope compositions of O_3, O_2, and CO_2 is limited. Observational data are needed urgently to confirm models for today's atmosphere, especially the upper atmo-sphere compositions. These efforts will help to improve our ability to recon-struct paleo-atmospheric conditions based on proxies such as sulfate $\Delta^{17}O$.

Further developments could be made if the analytical accuracy on the $\Delta^{17}O$ could be improved from today's ±0.005‰ to ± 0.001‰ (s.e.m.), or entirely new methods of measurement were identified (Eiler et al., 2017; Rumble, 2018).

9. Key Papers

Theories

Matsuhisa, Y., Goldsmith, J. R., and Clayton, R. N., 1978, Mechanisms of hydrothermal crystallization of quartz at 250°C and 15 kilobars.

Geochimica et Cosmochimica Acta, v. 42, no. 2, pp. 173–182. Although the title sounds like marginally relevant research on triple oxygen isotopes, two important original contributions were made in this paper. This was the first attempt to calculate the exact triple oxygen isotope exponent values for both equilibrium and kinetic processes, as well as θ's relationship with temperature. It is humbling to see the clarity in the authors' conceptual understanding of the triple isotope system 40 years ago when they first explored this subject. For example, from the Appendix I, "Since the fractionation ratios, both for equilibrium and for kinetic processes, depend somewhat on the particular molecules or crystals involved, it must be expected that an unsystematic sampling of a large and complicated system, such as the surface of the Earth, will yield a slope on a graph of $\delta^{17}O$ vs. $\delta^{18}O$ which is some average of the many individual slopes for specific processes." Second, the well-known "three-isotope method" for experimentally calibrating equilibrium isotope fractionation factors was first proposed and exercised in this paper. The "three-isotope method" has recently been increasingly used in determining equilibrium isotope fractionation factors for many new metal elements because of improved analytical capability and new instruments such as the Multicollector-Inductively Coupled Plasma Mass Spectrometer (MC-ICPMS). Throughout this 1978 paper, however, the authors repeatedly cautioned its application. In fact, much of the conclusions covered the three-isotope-method's utility in detecting disequilibrium.

Gao, Y. Q., and Marcus, R. A., 2001, Strange and unconventional isotope effects in ozone formation. *Science*, v. 293, no. 5528, pp. 259–263. Both mass-dependent and non-mass-dependent isotope enrichment in O_3 formation were treated based on RRKM (Rice, Ramsperger, Kassel, Marcus) theory for a bimolecular recombination reaction. The NMD ^{17}O enrichment is sensitive to an η-factor, a consequence of molecular symmetry.

Young, E. D., Galy, A., and Nagahara, H., 2002, Kinetic and equilibrium mass-dependent isotope fractionation laws in nature and their geochemical and cosmochemical significance. *Geochimica et Cosmochimica Acta*, v. 66, no. 6, pp. 1095–1104. This highly cited theoretical study was partially motivated by the concern that the small negative $\Delta^{17}O$ value observed in today's near-surface O_2 may be a result of θ differences for different mass-dependent processes rather than some mixing of ^{17}O-depleted O_2 derived from the stratosphere.

Cao, X. B., and Liu, Y., 2011, Equilibrium mass-dependent fractionation relationships for triple oxygen isotopes. *Geochimica et Cosmochimica Acta*, v. 75, no. 23, pp. 7435–7445. It is the first high-precision theoretical computation of equilibrium triple isotope relationship θ and its variation with temperature. The θ value is a function of α, β (i.e., the fractionation factor between an oxygen-bearing compound and the mono-atomic O), and a new concept κ ($\equiv \ln^{17}\beta/\ln^{18}\beta$ in the case of triple oxygen).

Bao, H. M., Cao, X. B., and Hayles, J. A., 2015, The confines of triple oxygen isotope exponents in elemental and complex mass-dependent processes. *Geochimica et Cosmochimica Acta*, v. 170, pp. 39–50. Much of the conceptual framework on the triple isotope relationship reviewed here, ranging from fundamental concepts to their manifestation in complex processes, came from this paper.

Original discoveries and applications

Clayton, R. N., Grossman, L., and Mayeda, T. K., 1973, A component of primitive nuclear composition in carbonaceous chondrites. *Science*, v. 182, pp. 485–488. This is the study I referred to in Section 2 above, "The first measurement of $\delta^{17}O$ value."

Thiemens, M. H., and Heidenreich, J. E., III, 1983, The mass-independent fractionation of oxygen: a novel isotope effect and its possible cosmochemical implications. *Science*, v. 219, no. 4588, pp. 1073–1075. This study first reported that O_3 produced by electrical discharge of O_2 in the laboratory can have large NMD ^{17}O anomalies.

Luz, B., Barkan, E., Bender, M. L., Thiemens, M. H., and Boering, K. A., 1999, Triple-isotope composition of atmospheric oxygen as a tracer of biosphere productivity. *Nature*, v. 400, no. 6744, pp. 547–550. Although there were unpublished reports by laboratories indicating that the $\delta^{17}O$ of air O_2 is slightly off from that predicted by its corresponding $\delta^{18}O$ value assuming mass-dependent fractionation. This study is the first definite confirmation of this deviation, that is, anomalously depleted in ^{17}O relative to SMOW. The authors attributed the cause of the NMD depletion to stratospheric O_3-CO_2-O_2 reaction network and demonstrated that O_2's $\Delta^{17}O$ value can be used as a proxy for biological primary productivity or photosynthetic O_2 production and mixing rates.

Bao, H., Lyons, J. R., and Zhou, C., 2008, Triple oxygen isotope evidence for elevated CO_2 levels after a Neoproterozoic glaciation. *Nature*, v. 453, no. 7194, pp. 504–506. While Bao et al. (2000b) first reported large NMD ^{17}O anomalies in the rock record, mainly in sulfate minerals, these anomalies were all enrichment of ^{17}O, that is, $\Delta^{17}O > 0$. They were mostly discovered in arid or semi-arid environments and relatively young, younger than 35 Ma in age. This study is the first to report geological minerals that have the $\Delta^{17}O < 0$. These large degrees of NMD ^{17}O depletion occurred in the immediate aftermath of Marinoan snowball Earth at ~635 Ma. With other constraints, this study linked the anomalous ^{17}O depletion in sulfate oxygen to that in near-surface air O_2, with the most probable cause being an ultra-high CO_2 concentration, hundreds of times that of today's level, at the time of deposition, thus providing arguably the strongest support to the snowball Earth hypothesis (Hoffman et al., 1998).

Barkan, E., and Luz, B., 2007, Diffusivity fractionations of $H_2{}^{16}O/H_2$ ${}^{17}O$ and $H_2{}^{16}O/H_2{}^{18}O$ in air and their implications for isotope hydrology. *Rapid Communications in Mass Spectrometry*, v. 21, no. 18, pp. 2999–3005. The first high-precision experimental measurement of θ value for water vapor diffusion in air, demonstrating that H_2O liquid-vapor equilibrium and vapor diffusion have different θ values. This study remains one of the very few that calibrated θ values for elementary processes.

Review papers

Thiemens, M. H., 2006, History and applications of mass-independent isotope effects. *Annual Review of Earth and Planetary Sciences*, v. 34, pp. 217–262. The origin and applications of NMD oxygen isotope signatures are well covered for cases pertinent to recent atmospheric chemistry and planetary sciences.

Bao, H. M., Cao, X. B., and Hayles, J. A., 2016, Triple Oxygen Isotopes: Fundamental Relationships and Applications, in Jeanloz, R., and Freeman, K. H. (Eds.), *Annual Review of Earth and Planetary Sciences*, Volume 44: Palo Alto, Annual Reviews, pp. 463–492. Basic concepts, laboratory measurements, and applications of triple oxygen isotopes as of 2015 were critically reviewed and synthesized in this paper. One key emphasis is the distinction of the concepts of θ, S, and C in the terminology of the triple isotope research community. The distinction aimed at clarifying confusions still plaguing the community today.

Dauphas, N. and Schauble, E. A., 2016, Mass Fractionation Laws, Mass-Independent Effects, and Isotopic Anomalies, in Jeanloz, R., and Freeman, K. H. (Eds.), *Annual Review of Earth and Planetary Sciences*, Volume 44: Palo Alto, Annual Reviews, pp. 709–783. This paper examines the origin of mass-anomalous isotope signatures from a broad perspective, including kinetic versus equilibrium, photolysis, photodissociation, magnetic, nuclear field shift, and nucleosynthetic effects.

References

Aston, F.W. (1919a) The Constitution of the Elements. *Nature* 104, 393.

Aston, F.W. (1919b) Neon. *Nature* 104, 334.

Aston, F.W. (1919c) A positive ray spectrograph. *Philosophical Magazine* 38, 707–714.

Aston, F.W. (1920a) The constitution of atmospheric neon. *Philosophical Magazine* 39, 449–455.

Aston, F.W. (1920b) Isotopes and Atomic Weights. *Nature* 105, 617.

Aston, F.W. (1929) The Constitution of Oxygen. *Nature* 123, 488.

Aston, F.W. (1932) Mass-Spectra of Helium and Oxygen. *Nature* 130, 21.

Babikov, D., Kendrick, B.K., Walker, R.B., Pack, R.T., Fleurat-Lesard, P. and Schinke, R. (2003) Formation of ozone: Metastable states and anomalous isotope effect. *Journal of Chemical Physics* 119, 2577–2589.

Bao, H. (2005) Sulfate in modern playa settings and in ash beds in hyperarid deserts: Implication on the origin of ^{17}O-anomalous sulfate in an Oligocene ash bed. *Chemical Geology* 214, 127–134.

Bao, H. (2015) Sulfate: A time capsule for Earth's O_2, O_3, and H_2O. *Chemical Geology* 395, 108–118.

Bao, H., Campbell, D.A., Bockheim, J.G. and Thiemens, M.H. (2000a) Origins of sulphate in Antarctic dry-valley soils as deduced from anomalous O-17 compositions. *Nature* 407, 499–502.

Bao, H. M., Cao, X. B., and Hayles, J. A. (2016) Triple Oxygen Isotopes: Fundamental Relationships and Applications, in: Jeanloz, R. and Freeman K. H. (Eds.), *Annual Review of Earth and Planetary Sciences*, Volume 44: Palo Alto, Annual Reviews, pp. 463–492.

Bao, H., Chen, Z.-Q. and Zhou, C. (2012) An ^{17}O record of late Neoproterozoic glaciation in the Kimberley region, Western Australia. *Precambrian Research* 216–219, 152–161.

Bao, H., Fairchild, I.J., Wynn, P.M. and Spoetl, C. (2009) Stretching the Envelope of Past Surface Environments: Neoproterozoic Glacial Lakes from Svalbard. *Science* 323, 119–122.

Bao, H. and Gu, B.H. (2004) Natural perchlorate has a unique oxygen isotope signature. *Environmental Science & Technology* 38, 5073–5077.

Bao, H., Lyons, J. R., and Zhou, C. (2008) Triple oxygen isotope evidence for elevated CO_2 levels after a Neoproterozoic glaciation. *Nature* 453(7194), 504–506.

Bao, H. and Marchant, D.R. (2006) Quantifying sulfate components and their variations in soils of the McMurdo Dry Valleys, Antarctica. *Journal of Geophysical Research-Atmospheres* 111.

Bao, H. and Thiemens, M.H. (2000) Generation of O_2 from $BaSO_4$ using a CO_2-laser fluorination system for simultaneous analysis of $\delta^{18}O$ and $\delta^{17}O$. *Analytical Chemistry* 72, 4029–4032.

Bao, H., Thiemens, M.H., Farquhar, J., Campbell, D.A., Lee, C.C.W., Heine, K. and Loope, D.B. (2000b) Anomalous ^{17}O compositions in massive sulphate deposits on the Earth. *Nature* 406, 176–178.

Bao, H.M., Thiemens, M.H. and Heine, K. (2001) Oxygen-17 excesses of the Central Namib gypcretes: spatial distribution. *Earth and Planetary Science Letters* 192, 125–135.

Bao, H., Thiemens, M.H., Loope, D.B. and Yuan, X.L. (2003) Sulfate oxygen-17 anomaly in an Oligocene ash bed in mid-North America: Was it the dry fogs? *Geophysical Research Letters* 30.

Bao, H., Yu, S.C. and Tong, D.Q. (2010) Massive volcanic SO_2 oxidation and sulphate aerosol deposition in Cenozoic North America. *Nature* 465, 909–912.

Barkan, E. and Luz, B. (2007) Diffusivity fractionations of $H_2^{16}O/H_2^{17}O$ and $H_2^{16}O/H_2^{18}O$ in air and their implications for isotope hydrology. *Rapid Communications in Mass Spectrometry* 21(18), 2999–3005.

Barkan, E. and Luz, B. (2012) High-precision measurements of $^{17}O/^{16}O$ and $^{18}O/^{16}O$ ratios in CO_2. *Rapid Communications in Mass Spectrometry* 26, 2733–2738.

Benn, D.I., Le Hir, G., Bao, H.M., Donnadieu, Y., Dumas, C., Fleming, E.J., Hambrey, M.J., McMillan, E.A., Petronis, M.S., Ramstein, G., Stevenson, C. T.E., Wynn, P.M. and Fairchild, I.J. (2015) Orbitally forced ice sheet fluctuations during the Marinoan Snowball Earth glaciation. *Nature Geoscience* 8, 704–+.

Bigeleisen, J. (2006) Theoretical Basis of Isotope Effects from an Autobiographical Perspective, in: Kohen, A., Limbach, H.-H. (Eds.), *Isotope Effects in Chemistry and Biology*. Taylor & Francis Group, LLC, Boca Raton London New York, pp. 1–39.

Bigeleisen, J., Mayer, M. G. (1947) Calculation of equilibrium constants for isotopic exchange reactions. *Journal of Chemical Physics* 15, 261–267.

Bindeman, I.N., Zakharov, D.O., Palandri, J., Greber, N.D., Dauphas, N., Retallack, G.J., Hofmann, A., Lackey, J.S. and Bekker, A. (2018) Rapid emergence of subaerial landmasses and onset of a modern hydrologic cycle 2.5 billion years ago. *Nature* 557, 545–548.

Blackett, P.M.S. (1925) The ejection of protons from nitrogen nuclei, photographed by the Wilson method. *Proceedings of the Royal Society of London. Series A* 107, 349–360.

Brickwedde, F.G. (1982) Urey, Harold and the discovery of deuterium. *Physics Today* 35, 34–39.

Cao, X.B. and Bao, H.M. (2013) Dynamic model constraints on oxygen-17 depletion in atmospheric O_2 after a snowball Earth. *Proceedings of the National Academy of Sciences of the United States of America* 110, 14546–14550.

Clayton, R.N. (2007) Isotopes: From Earth to the Solar System. *Annual Review of Earth and Planetary Sciences* 35, 1–19.

Clayton, R.N. (2008) Oxygen isotopes in the early Solar System – A historical perspective, in: MacPherson, G.J., Mittlefehldt, D.W., Jones, J.H., Simon, S. B. (Eds.), *Oxygen in the Solar System*, pp. 5–14. Mineralogical Society of America, Washington, D.C.

Clayton, R.N. and Mayeda, T.K. (1984) The oxygen isotope record in Murchison and other carbonaceous chondrites. *Earth and Planetary Science Letters* 67, 151–161.

Clayton, R.N., Grossman, L. and Mayeda, T.K. (1973) A component of primitive nuclear composition in carbonaceous chondrites. *Science* 182, 485–488.

Craig, H. (1957) Isotopic standards for carbon and oxygen and correction factors for mass-spectrometric analysis of carbon dioxide. *Geochimica et Cosmochimica Acta* 12, 133–149.

Craig, H. (1961) Standard for reporting concentrations of deuterium and oxygen-18 in natural waters. *Science* 133, 1833–1834.

Crockford, P.W., Cowie, B.R., Johnston, D.T., Hoffman, P.F., Sugiyama, I., Pellerin, A., Bui, T.H., Hayles, J., Halverson, G.P., Macdonald, F.A. and Wing, B.A. (2016) Triple oxygen and multiple sulfur isotope constraints on the evolution of the post-Marinoan sulfur cycle. *Earth and Planetary Science Letters* 435, 74–83.

Crockford, P.W., Hayles, J.A., Bao, H.M., Planavsky, N.J., Bekker, A., Fralick, P.W., Halverson, G.P., Bui, T.H., Peng, Y.B. and Wing, B.A. (2018a) Triple oxygen isotope evidence for limited mid-Proterozoic primary productivity. *Nature* 559, 613–+.

Crockford, P.W., Hodgskiss, M.S.W., Uhlein, G.J., Caxito, F., Hayles, J.A. and Halverson, G.P. (2018b) Linking paleocontinents through triple oxygen isotope anomalies. *Geology* 46, 179–182.

Dauphas, N. and Schauble, E. A. (2016) Mass Fractionation Laws, Mass-Independent Effects, and Isotopic Anomalies, in: Jeanloz, R. and Freeman, K. H. (Eds.), *Annual Review of Earth and Planetary Sciences*, Vol 44, Volume 44: Palo Alto, Annual Reviews, pp. 709–783.

Eiler, J., Cesar, J., Chimiak, L., Dallas, B., Grice, K., Griep-Raming, J., Juchelka, D., Kitchen, N., Lloyd, M., Makarov, A., Robins, R. and Schwieters, J. (2017) Analysis of molecular isotopic structures at high precision and accuracy by Orbitrap mass spectrometry. *International Journal of Mass Spectrometry* 422, 126–142.

Evans, N.P., Bauska, T. K., Gázquez, F., Curtis, J.H., Brenner, M., Hodell, D.A. (2018) Quantification of drought during the collapse of the classic Maya civilization. *Science* 361, 498–501.

Farquhar, J., Johnston, D.T., Wing, B.A., Habicht, K.S., Canfield, D.E., Airieau, S. and Thiemens, M.H. (2003) Multiple sulphur isotopic interpretations of biosynthetic pathways: implications for biological signatures in the sulphur isotope record. *Geobiology* 1, 27–36.

Gao, Y.Q. and Marcus, R.A. (2001) Strange and unconventional isotope effects in ozone formation. *Science* 293, 259–263.

Gao, Y.Q. and Marcus, R.A. (2002) On the theory of the strange and unconventional isotopic effects in ozone formation. *Journal of Chemical Physics* 116, 137–154.

Gázquez, F. Morellón M., Bauska, T., Herwartz, D., Surma, J., Moreno, A. Staubwasser, M., Valero-Garcés, B., Delgado-Huertas, A., Hodella, D.A. (2018) Triple oxygen and hydrogen isotopes of gypsum hydration water for quantitative paleo-humidity reconstruction. *Earth and Planetary Science Letters* 481, 177–188.

Giauque, W.F. and Johnston, H.L. (1929a) An Isotope of Oxygen, Mass 18. *Nature* 123, 318.

Giauque, W.F. and Johnston, H.L. (1929b) An Isotope of Oxygen of Mass 17 in the Earth's Atmosphere. *Nature* 123, 831.

Hayles, A., Cao, X. and Bao, H. (2017) The statistical mechanical basis of the triple isotope fractionation relationship. *Geochemical Perspectives Letters* 3, 1–11.

Hayles, J., Gao, C.H., Cao, X.B., Liu, Y. and Bao, H.M. (2018) Theoretical calibration of the triple oxygen isotope thermometer. *Geochimica et Cosmochimica Acta* 235, 237–245.

Herwartz, D., Pack, A., Krylov, D., Xiao, Y., Muehlenbachs, K., Sengupta, S. and Di Rocco, T. (2015) Revealing the climate of snowball Earth from $\Delta^{17}O$ systematics of hydrothermal rocks. *Proceedings of the National Academy of Sciences* 112, 5337–5341.

Herwartz, D., Surma, J., Voigt, C., Assonov, S. and Staubwasser, M. (2017) Triple oxygen isotope systematics of structurally bonded water in gypsum. *Geochimica Et Cosmochimica Acta* 209, 254–266.

Hoffman, P.F., Kaufman, A.J., Halverson, G.P. and Schrag, D.P. (1998) A Neoproterozoic snowball earth. *Science* 281, 1342–1346.

Hogness, T.R. and Kvalnes, H.M. (1928) Isotopes of neon. *Nature* 122, 441–441.

Holton, J.R., Haynes, P.H., McIntyre, M.E., Douglass, A.R., Rood, R.B. and Pfister, L. (1995) Stratosphere-troposphere exchange. *Reviews of Geophysics* 33, 403–439.

Hughes, J. (2009) Making isotopes matter: Francis Aston and the mass-spectrograph. *Dynamis* 29, 131–165.

Johnston, J.C. and Thiemens, M.H. (1997) The isotopic composition of tropospheric ozone in three environments. *Journal of Geophysical Research – Atmospheres* 102, 25395–25404.

Kawagucci, S., Tsunogai, U., Kudo, S., Nakagawa, F., Honda, H., Aoki, S., Nakazawa, T., Tsutsumi, M. and Gamo, T. (2008) Long-term observation of mass-independent oxygen isotope anomaly in stratospheric CO_2. *Atmospheric Chemistry and Physics* 8, 6189–6197.

Killingsworth, B.A., Bao, H.M. and Kohl, I.E. (2018) Assessing Pyrite-Derived Sulfate in the Mississippi River with Four Years of Sulfur and Triple-Oxygen Isotope Data. *Environmental Science & Technology* 52, 6126–6136.

Killingsworth, B.A., Hayles, J.A., Zhou, C.M. and Bao, H. (2013) Sedimentary constraints on the duration of the Marinoan Oxygen-17 Depletion (MOSD) event. *Proceedings of the National Academy of Sciences of the United States of America* 110, 17686–17690.

Krankowsky, D., Lammerzahl, P., Mauersberger, K., Janssen, C., Tuzson, B. and Rockmann, T. (2007) Stratospheric ozone isotope fractionations derived from collected samples. *Journal of Geophysical Research-Atmospheres* 112.

Levin, N.E., Raub, T.D., Dauphas, N. and Eiler, J.M. (2014) Triple oxygen isotope variations in sedimentary rocks. *Geochimica et Cosmochimica Acta* 139, 173–189.

Liang, M.C., Mahata, S., Laskar, A.H., Thiemens, M.H. and Newman, S. (2017) Oxygen isotope anomaly in tropospheric CO_2 and implications for CO_2 residence time in the atmosphere and gross primary productivity. *Scientific Reports* 7, 12.

Lyons, J.R. (2001) Transfer of mass-independent fractionation in ozone to other oxygen-containing radicals in the atmosphere. *Geophysical Research Letters* 28, 3231–3234.

Luz, B. and Barkan, E. (2010) Variations of $^{17}O/^{16}O$ and $^{18}O/^{16}O$ in meteoric waters. *Geochimica et Cosmochimica Acta* 74, 6276–6286.

Luz, B., Barkan, E., Bender, M. L., Thiemens, M. H. and Boering, K. A. (1999) Triple-isotope composition of atmospheric oxygen as a tracer of biosphere productivity. *Nature* 400, 547–550.

Mahata, S., Bhattacharya, S.K. and Liang, M.C. (2016) An improved method of high-precision determination of $\Delta^{17}O$ of CO_2 by catalyzed exchange with O_2 using hot platinum. *Rapid Communications in Mass Spectrometry* 30, 119–131.

Matsuhisa, Y., Goldsmith, J. R. and Clayton, R. N. (1978) Mechanisms of hydrothermal crystallization of quartz at 250°C and 15 kilobars. *Geochimica et Cosmochimica Acta* 42(2), 173–182.

Michalski, G., Savarino, J., Bohlke, J.K. and Thiemens, M. (2002) Determination of the total oxygen isotopic composition of nitrate and the calibration of a $\Delta^{17}O$ nitrate reference material. *Analytical Chemistry* 74, 4989–4993.

Michalski, G., Scott, Z., Kabiling, M. and Thiemens, M.H. (2003) First measurements and modeling of $\Delta^{17}O$ in atmospheric nitrate. *Geophysical Research Letters* 30, 1870; 1810.1029/2003GL017015.

Murphey, B.F. (1941) Relative abundances of the oxygen isotopes. *Physical Review* 59, 320–320.

Pack, A. and Herwartz, D. (2014) The triple oxygen isotope composition of the Earth mantle and understanding $\Delta^{17}O$ variations in terrestrial rocks and minerals. *Earth and Planetary Science Letters* 390, 138–145.

Passey, B.H., Hu, H.T., Ji, H.Y., Montanari, S., Li, S.N., Henkes, G.A. and Levin, N.E. (2014) Triple oxygen isotopes in biogenic and sedimentary carbonates. *Geochimica et Cosmochimica Acta* 141, 1–25.

Peng, Y.B., Bao, H., Zhou, C.M. and Yuan, X.L. (2011) [17]O-depleted barite from two Marinoan cap dolostone sections, South China. *Earth and Planetary Science Letters* 305, 21–31.

Rumble, D. (2018) The third isotope of the third element on the third planet. *American Mineralogist* 103, 1–10.

Savarino, J., Lee, C.C.W. and Thiemens, M.H. (2000) Laboratory oxygen isotopic study of sulfur (IV) oxidation: Origin of the mass-independent oxygen isotopic anomaly in atmospheric sulfates and sulfate mineral deposits on Earth. *Journal of Geophysical Research-Atmospheres* 105, 29079–29088.

Sharp, Z.D., Gibbons, J.A., Maltsev, O., Atudorei, V., Pack, A., Sengupta, S., Shock, E.L. and Knauth, L.P. (2016) A calibration of the triple oxygen isotope fractionation in the SiO_2–H_2O system and applications to natural samples. *Geochimica et Cosmochimica Acta* 186 105–119.

Sun, T., Bao, H.M., Reich, M. and Hemming, S.R. (2018) More than ten million years of hyper-aridity recorded in the Atacama Gravels. *Geochimica et Cosmochimica Acta* 227, 123–132.

Tarasicka, D.W., Carey-Smith, T.K., Hocking, W.K. Moeini, O., He, H., Liu, J., Osman, M.K., Thompson, A.M., Johnson, B.J., Oltmans, S.J. and Merrilli, J. T. (2019) Quantifying stratosphere-troposphere transport of ozone using balloon-borne ozonesondes, radar windprofilers and trajectory models. *Atmospheric Environment* 198, 496–509.

Thiemens, M.H. (1999) Mass-independent isotope effects in planetary atmospheres and the early solar system. *Science* 283, 341–345.

Thiemens, M.H., and Heidenreich, J. E., III (1983) The mass-independent fractionation of oxygen: a novel isotope effect and its possible cosmochemical implications: *Science* 219 (4588), 1073–1075.

Thiemens, M.H., Jackson, T., Zipf, E.C., Erdman, P.W. and Vanegmond, C. (1995) Carbon Dioxide and Oxygen Isotope Anomalies in the Mesosphere and Stratosphere. *Science* 270, 969–972.

Thompson, J.J. (1913) Bakerian Lecture: Rays of Positive Electricity. *Proceedings of the Royal Society A* 89, 1–20.

Urey, H.C. (1947) The thermodynamic properties of isotopic substances. *Journal of the Chemical Society* 562–581.

Vicars, W.C. and Savarino, J. (2014) Quantitative constraints on the O^{17}-excess ($\Delta^{17}O$) signature of surface ozone: Ambient measurements from 50 degrees N to 50 degrees S using the nitrite-coated filter technique. *Geochimica et Cosmochimica Acta* 135, 270–287.

Wostbrock, J.A.G., Sharp, Z.D., Sanchez-Yanez, C., Reich, M., van den Heuvel, D.B. and Benning, L.G. (2018) Calibration and application of silica-water triple oxygen isotope thermometry to geothermal systems in Iceland and Chile. *Geochimica et Cosmochimica Acta* 234, 84–97.

Young, E. D., Galy, A., and Nagahara, H. (2002) Kinetic and equilibrium mass-dependent isotope fractionation laws in nature and their geochemical and cosmochemical significance. *Geochimica et Cosmochimica Acta* 66(6), 1095–1104.

Yung, Y.L., Lee, A.Y.T., Irion, F.W., DeMore, W.B. and Wen, J. (1997) Carbon dioxide in the atmosphere: Isotopic exchange with ozone and its use as a tracer in the middle atmosphere. *Journal of Geophysical Research-Atmospheres* 102, 10857–10866.

Acknowledgement

Financial support is partially provided by the strategic priority research program (B) of the Chinese Academy of Sciences (XDB18010104), China NNSF Grant 41490635, Short-term QianRen program, and Charles L. Jones professorship fund.

Cambridge Elements ≡

Elements in Geochemical Tracers in Earth System Science

Timothy Lyons
University of California
Timothy Lyons is a Distinguished Professor of Biogeochemistry in the Department of Earth Sciences at the University of California, Riverside. He is an expert in the use of geochemical tracers for applications in astrobiology, geobiology, and Earth history. Professor Lyons leads the "Alternative Earths" team of the NASA Astrobiology Institute and the Alternative Earths Astrobiology Center at UC Riverside.

Alexandra Turchyn
University of Cambridge
Alexandra Turchyn is a University Reader in Biogeochemistry in the Department of Earth Sciences at the University of Cambridge. Her primary research interests are in isotope geochemistry and the application of geochemistry to interrogate modern and past environments.

Chris Reinhard
Georgia Institute of Technology
Chris Reinhard is an Assistant Professor in the Department of Earth and Atmospheric Sciences at the Georgia Institute of Technology. His research focuses on biogeochemistry and paleoclimatology, and he is an Institutional PI on the "Alternative Earths" team of the NASA Astrobiology Institute.

About the series
This innovative series provides authoritative, concise overviews of the many novel isotope and elemental systems that can be used as 'proxies' or 'geochemical tracers' to reconstruct past environments over thousands to millions to billions of years – from the evolving chemistry of the atmosphere and oceans to their cause-and-effect relationships with life.
Covering a wide variety of geochemical tracers, the series reviews each method in terms of the geochemical underpinnings, the promises and pitfalls, and the "state-of-the-art" and future prospects, providing a dynamic reference for graduate students and researchers in geochemistry, astrobiology, paleontology, paleoceanography, and paleoclimatology.
The short, timely, broadly accessible Elements provide much-needed primers for a wide audience – highlighting the cutting-edge of both new and established proxies as applied to diverse questions about Earth system evolution over wide-ranging time scales.

Cambridge Elements ☰

Elements in Geochemical Tracers in Earth System Science

Elements in the series

The Uranium Isotope Paleoredox Proxy
Kimberly V. Lau et al.

Triple Oxygen Isotopes
Huiming Bao